FIELD TO FARM

D.Acreman

First published in Great Britain in 2009

Bulldozer Publishing
49 Courtlands Drive, Watford, herts.
fieldtofarm@hotmail.co.uk

ISBN 978-0-9561956-0-9

D.Acreman has asserted his right to be identified as the author
of this work under the Copyright, Designs and Patents Act 1988

All rights reserved. No part of this publication may be reproduced, In any material form (including photocopying or storing in any medium by electronic means and whether or not transiently or incidentally to some other use of this publication) without the written permission of the copyright owner except in accordance with the provisions of the Copyright, Designs and Patent Act 1988.All matters relating to copyright should be addressed to the publishers at the address above.

Whilst every care has been taken to ensure the accuracy of the content of this work, no responsibility for loss occasioned to any person acting or refraining from action as a result of the material in this publication can be accepted by the author or by the publisher. Appropriate professional advice should be sought before entering into any contractual obligations or financial commitments.

CONTENTS

INTRODUCTION	5
THE LAND	9
PLANNING	19
THE DWELLING	24
THE MEASOR PRINCIPLE	25
CONSTRUCTION	27
THE EXCAVATIONS	28
THE BUILDING	32
THE SERVICES	36
THE BUSINESS	37
THE TESTS	39
MAIN ANIMAL ENTERPRISES	40
UNSUITABLE MAIN ANIMAL ENTERPRISES	45
ADD ON ANIMAL ENTERPRISES	46
HORSES	49
GROWING	50
TOURING CARAVAN SITE	53
SKILLS	54
LEGISLATION	55
RELEVENT SECTIONS OF PLANNING LAW	56

INTRODUCTION

It does seem unlikely in our heavily regulated planning system, that it would be possible to carry out any development on agricultural land without planning permission. However as this book will demonstrate it is possible, to live, build a barn and start a business, all without planning permission, on ordinary agricultural land in the United Kingdom

This book uses current planning law only and does not rely on using the Human Rights Act, used by others to occupy land.

Most people at one time or another would like to get away from it all, to escape the rat race, live the "good life". To buy a piece of land, live on it and keep a few animals. If you are one of these people and this is the time, then this book is for you. It will guide you through the process, explaining all you need to know, pointing out all the pitfalls. The guide will cover the following.

1) Acquisition of the land, what to look for what to avoid.
2) Explain why planning permission is not required.
3) The process of siting and construction of the 5000 sq ft barn.
4) Selection and siting of the dwelling.
5) The infrastructure and services.
6) A selection of businesses to run and the tests that will need to be met to get planning permission for a house.
7) How to set up and run your own touring caravan park without planning permission.

The usual way to start a smallholding is to buy a piece of land, (you will be very lucky to find a piece close to where you live) then make a planning application to live on the land. For this application to be successful you will be required to prove functional need, which simply means to justify why you need to live on the land. This always means care and welfare of agricultural animals. This can be problematic if you have no animals. You can only put in a business plan supported by information which you will need to get from an agricultural planning consultant, which costs money and time. The most successful method is to have an animal enterprise up and running and then apply for planning permission, although this can have its own problems. The council may well take the view that you have operated the animal enterprise without living on the land, so what is the need to live on it now, A "catch 22" situation. This approach can take a great deal of time and money, particularly if you have already bought the land. You will receive no help from the Local Authority planning department. Generally you will be fed misinformation about what you can and cannot do on agricultural land. If you are successful in obtaining planning it will be for a mobile home for a period of 3 years (no extensions) and in this period you will need to prove viability. The test is earning the minimum wage for a period of 1 year out of the 3. If this test is met then planning permission would be granted for a permanent dwelling which would be of a size which would be commensurate with the income that can be earned from the holding, on the minimum wage.

It would be a small house. It will also be necessary to demonstrate that the business is sustainable; this is normally demonstrated by an additional investment in the business or in an additional enterprise or building.

The house is likely to come with some planning restriction placed up on it. The usual one is that the house and land can only be sold together or an agricultural tie which restricts the people who can live in the house. Usually only people who are working in agriculture or forestry or who have retired from same are allowed. Sometimes some of the permitted development rights are removed which means you can not extend the house without planning permission and this is unlikely to be granted. They may also remove the right to have a mobile home within the curtilage of a dwelling house I am not sure that this would be enforceable if the measor principle is applied, which is explained further in the planning chapter on the mobile home. If you use this book as a guide to set up your smallholding, the outcome will be more positive, no costly planning application, only a 28 day notice (an agricultural prior notice consent form). You may move onto the land and live on it for up to 5 years without making a planning application, this gives ample time to establish an animal based business without the "catch 22" situation, so any land that is bought can, with certainty be used as a smallholding. The work to set up the smallholding could start in weeks not months, or years, this now gives you the time to become a viable profitable enterprise rather than having to be profitable in your 1 year in 3. Some people believe that being self sufficient is the way they want to run their smallholding, a, laudable ideal however this will be a difficult way to get planning permission for a house. There are Government guidelines given to planners, whereby some smallholdings can operate at a subsistence level, which is allowed so long as they contribute in other ways, an example would be enhancing and maintaining the landscape. This would be a difficult rout to getting planning permission, the house you would be allowed to build would be tiny. It will be far easier to follow the system set out in this guide with a more certain outcome. To do this you will need to make money as much money as you can. The more income you have, the better the house you can build. After you have built your house you can be as self sufficient as you like.

The only way this system will work is to be realistic. Forget the good life, every opportunity for making money must be exploited. That is why it is critical to buy the right piece of land in the right place and to diversify. You will need a main animal enterprise to satisfy planning but you can also do lots of other things at the same time. You could have a petting zoo and a touring caravan park and do seasonal things such as sell bedding plants in spring a maize maze in summer pumpkins in October, Christmas trees and turkeys or even a farm shop. The possibilities are endless. The one thing that I will keep coming back to is the land. It must be right for what you want to do with it. Flat at least in part, if you want a touring caravan park, not wet or heavy soil if you want to grow Christmas trees. It is better to read this book all the way through, decide what businesses you would like to run and then see if the land is suitable. Do your local research to see if there are opportunities for what you want to do. If not change what you want to do or change the location. Whatever land you select it must have road frontage. The road must go somewhere and must have passing traffic (as much as possible). Without this traffic you have little chance of selling things to the public without spending a lot of money on advertising. For a relatively modest investment you should end up with a smallholding with an asset value worth at least two or three times your outlay it could be even more, so from an investment point of view it would be hard to beat.

The qualities necessary to create a smallholding are: some degree of physical fitness as there is quite a lot of manual work involved, if you are not fit to start with you soon will be! Common sense, if you do not know ask. If you decide to keep a particular animal take a course. These are normally run by the people who specialise in that animal and sell them, you can buy books get magazines on smallholdings. You will soon learn, everybody has to start somewhere.

Being tenacious without being stubborn, stick with it but not to the bitter end. If something is not working, change it. If it looks like it's not going to work, drop it. Always try to be adaptable. Watch, listen and learn. There is always another way to do any job and you never know, it might be better than yours. If you are married you must both be committed to doing this. It's a hard slog and you are going to need each other, if both your hearts are not in it, don't do it. Always remember that "The Good Life" was not only a joke about Margo and Jerry the biggest joke was the Good's and their fixed ideas and intransigent ways of carrying out their ridiculous plans. So please be a realist and this method will work for you. Now you can get on with reading how to do it, good luck!

THE LAND

Buy the biggest piece of land that you can afford. You will probably never get the opportunity to buy a piece of land adjoining yours in your life time. If you ever need to expand your smallholding you have the land to do so. Remember what Mark Twain said "land is the best investment, it's the one thing they aren't making any more of" and he was right.
The land that we need is ordinary agricultural land, it can be called various names when it is being advertised green belt, accommodation land, permanent pasture and grazing land; the list goes on and on. If a piece of land is marketed as arable land, particularly if it is graded it is likely to be more expensive if it's in the right place and you can afford it, buy it. It is also likely to be ploughed it can always be put down to grass, but this will cost money and time (that is the only drawback) the upside is you will grow better grass and more of it. Stay away from pony paddocks, too expensive and too small, you sell surplus land as pony paddocks, not buy them.

We will require for planning purposes a minimum of 5 hectares (12.355 acres) of land, nothing smaller will do. You may have to buy a much larger piece, if that is what is available and if there is a local market for small pieces of land, sell some of it, but check before buying. Larger acreages are cheaper up to a certain size, cheaper than smaller pieces. The reason for this is that if a farmer wanted to buy the land and he didn't own any land nearby, he would have to transport the animals to and from the land, which isn't financially viable. However if the land adjoins his land he will pay almost any price to buy it. It does not work like this with arable land. If for example a block of 20 acres came up for sale, it would not be expensive to send along a tractor to seed, spray or harvest a remote field on the few occasions this would be necessary. We are after the bits of land in between too big for pony paddocks and too small for farmers.

LOCATION

This is the most difficult part in the search for land and also the most important, I will say this many times in this book I cannot emphasise enough that location is everything and if its right then success will follow, if its wrong the system may not succeed as well as it should or could even fail. You may be lucky in your search and find the perfect piece of land exactly where you want to live, not likely but possible. If you live in the southeast or Dorset or around any major conurbation most land for sale will be small pieces too small for our needs, normally in lots of approx 2-3 hectares (5 acres). The larger pieces are bought by building and property companies in the hope that one day they will get planning for housing. This practice has large tracts of land laying idle, used as dumping grounds as the companies owning it don't use it or let anybody use it, this practice should be stopped but that is unlikely to happen. On a brighter note other areas of the country will have land for sale Devon, West Wales, East Anglia to name but a few. These areas seem to have a history of either smallholding or breaking up farms so land in small parcels is plentiful. There are many other areas where land is for sale I have just highlighted the areas with plenty. I have even seen some for sale in Kent and reasonably inexpensive. So look in all areas never assume its not there. Look for it and you may get lucky. I am unable to say what a good price is for land, there are so many factors involved which will determine the lands value. The land is worth what you are prepared to pay for it, you should look at what other pieces of land have sold for in the area to get some idea of the going rate.

TOPOGRAPHY

The ideal piece of land would be flat with views, which is not very likely. What is more likely is a mixture of sloping and flat land. How sloping, only you know what is acceptable, after all you will be the one walking up and down the slope normally carrying something heavy. So think carefully! The flat part of the land will need to be near the road, although not too close, on this flat area the barn, the mobile home and the caravan site would be located, so it would need to be at least an acre. The land must be on a main road or a road that has good traffic flows. The land must front the road so that you can put up signs for things you are selling or activities on the land. Do not buy land that is not on the road, it will cause all sorts of problems and expense in the long run. Location is everything, if it is not in the right place, don't buy it. A good compromise with the land would be to have some flat some sloping and a copse or small wooded area, but not woodland, as planning is almost impossible. The road should be at least the size of a B road which should allow an articulated lorry to turn in from it without cutting too much of the corner. This would be for various deliveries including the barn the mobile home etc. The existing access on to the land should ideally be in a safe place, unlikely to be a hazard to any road users. It should be wide enough to take a lorry, with a 12 foot wide mobile on its back. If its not, it's not easy to move a field entrance, particularly on a busy road. The size and position of the access could affect any future planning applications particularly regarding any enlargement of the caravan site.

TO BE AVOIDED

COVENANTS

These are restrictions placed on the deeds prohibiting an activity on the land. It could for instance be no pigs or no caravans or even no house it is almost impossible to get them lifted and can cost a lot of money to try.

CLAWBACKS, UPLIFTS and OVERAGE.

These are all clauses, different names for the same thing. It means that in the event that planning permission is granted, even for one house, you will pay a percentage of the uplift in value, which goes to the seller of the land. This can be as high as 50% and the clawback can last for up to 50 years. If the seller is so worried about losing any future potential profit, he should keep the land. Don't ever buy land with this type of restriction placed upon it, the value of the land will be affected if you decide to sell.

ARTICLE 4 DIRECTION.

This is a restriction placed on the land by the council sanctioned by the secretary of state removing permitted development rights. A full explanation is given at the end of the planning chapter.

FISHING and SHOOTING RIGHTS (SPORTING RIGHTS)

These rights should be sold with the land. If they have been sold off separately then do not buy the land. They give a right of access without any redress to the owner of the land, it raises all sorts of security issues and personally. I wouldn't feel as though I owned the land fully.

MINERAL RIGHTS

These should be sold with the land as they allow the owner of the rights to extract minerals from beneath the land although access would have to be negotiated with the owner of the land. Always take legal advice as the wording of each agreement is different so don't get caught out.

RIGHTS of WAY and BRIDLEWAYS.

This is a tricky one. If you are a horse lover, being on a bridleway is perfect. If not, any public right of way across the land can cause all sorts of problems particularly lack of privacy and security. If a path crosses your land, any user of that right of way can sue the owner of the land if they are hurt whilst using it. Also these kind and helpful people have a tendency to feed the animals, normally with inappropriate food which can damage them and leave the bags they brought it in as litter. Think long and hard before making any decision.

ANCIENT MONUMENT

If any part of the land has anything which would be regarded as an ancient monument (these are always marked on the ordnance map, sometimes marked as tumulus if it's a burial site or similar). Do not buy this land as any site of this nature will have its permitted development rights removed and therefore useless for the method of planning that is to be used.

PLACES TO AVOID

There are designated areas in the UK where it is unlikely you will obtain planning permission the reason for this is normally preservation of landscape or protection of habitat of flora or fauna. I have listed these areas below always check with the council if the land you are interested in, is in a designated area, if it is don't buy it.

1) Land in National Parks or areas like the Norfolk and Suffolk Broads
2) Areas of Outstanding Natural Beauty
3) Sites of Special Scientific Interest (SSSI)
4) Woodland

To Comply with Planning
1) Land size minimum 5 hectares
2) Land not within 400 metres of a conservation area
3) Land not within 3 kilometres of the perimeter of an Airfield
4) Building area to be no larger than 465 square metres
5) Building to be no higher than 3 metres within 3 kilometres of the perimeter of an Airfield
6) Building to be no higher than 10 metres if more than 3 kilometres away from the Perimeter of an airfield

Agreement with the planning officer on the position of building on the land and the Colour of the cladding to be used.

If you want a touring caravan site try to be a fair distance from the next one, say 5 kilometres or so, unless you are in a high demand area. Research will be necessary to determine the level of demand locally.

THINGS TO LOOK FOR

If you use a local solicitor who carries out land purchases they will know what searches to carry out however look out for the list below.

1) Mine workings under the land.

2) Pylons passing over the land, this is fine if passing over a small part of the land but not near where you want to live or anywhere near it, this also applies to the cables.

3) Land with river or stream, lake or pond or any of the same nearby should be checked with the environment agency for potential flood risk.

4) Land adjacent to any industrial type building.

5) Land near a track used for off road vehicles or paintball events.

6) In some areas a lot of land was owned by the Crown or large estates over the years. They have sold off some of the land but retained the ribbons of land along the sides of roads. These are called ransom strips, they will not affect existing rights of way over them but you would have to pay if you wanted a new entrance onto your land.

SINGLE FARM PAYMENTS

These are paid by the Government according to the size of the land to the land owner. Sometimes the vendor will want reimbursement for this payment, it maybe necessary to register the land and it will certainly be necessary to apply for it annually. The agent will have all the details relating to that particular piece of land and if it is in a stewardship scheme.

Sometimes a farmer will sell land with standing crops. It would be easier to allow the farmer to harvest the crops themselves and put the land back to grass. All this would have to be negotiated. Usually your solicitor can tie this up legally but remember depending on the time of year you are purchasing you could lose months before you can get onto the land.

When you have located a piece of land that may be suitable at the right price, ask the agent about the land, do not just put it in the hands of your solicitor or this will cost you time and money. You do not involve a solicitor until it is necessary. If you ask the agent if the land has any covenants on it, he should be able to get the answers from the vendor. He may say it will show up on the searches It will, but this would be a long way down the line and will cost money. You will have used your solicitor, in which case if you abort the sale you will still have to pay them. Also you lose time because solicitors write to each other and it would take time to get the full information on a piece of land. While all this is going on you could be missing out on another piece of land. You must make your own inquiries. You certainly want to know from the agent the situation on covenants, sporting rights, rights of way in an area of outstanding natural beauty etc. Other information can be found from different sources, the Local Authority Planning department for an article 4 direction, an ordnance survey map of the area (the Explorer maps at 25 000 scale are the best) the map will tell you if the land is in a national park, indicate rights of way proximity of airfields etc. Only if all the answers you get back are positive do you then put it with your solicitor.

FINDING the LAND.

Looking for land on the internet is very frustrating if you put in the word "agent" you get all sorts of lists of estate agents. Not their web sites, just lists. If you go down the page don't bother with these look for the actual name of an estate agent, not sites saying they have all the agent site in an area. Firstly they haven't and secondly, there will be no web site link, this is a waste of time. Unless you see a name with a hyperlink (one that is coloured which can be clicked on and will take you to a website). I try to find estate agents with names that are old fashioned. These are usually old established a firms more likely to sell land.

Rightmove can be useful you can narrow down your area and there is a tab for land but this is usually for building plots. At least you can access the sites of agents in your search area and see what they do

If you look up Auctioneers, these companies are usually estate agents as well, particularly livestock or antique Auctioneers, they could also sell land.

Do not sign up to sites on the internet that say they have land for sale. Normally they don't and if they do, why would you pay a fee to be shown a piece of land. It certainly won't be cheap, it will be over priced or it would be on the open market leave these sites alone.

If you put into the search engine "land for sale" you will get all sorts of sites selling land but normally in small pieces. They buy a field and mark out roads with pegs and divide the field up into plots for houses as an investment with the hope of getting planning permission. None of these imaginary estates have ever got planning permission, as I am sure that if they did they would use it in their advertising. The only reason I am mentioning these companies is that they also offer land for (sale rarely stating the price) and if they do it is top dollar for rough land. These companies are to be avoided at your peril.

Whoever you deal with when you buy land do not tell them what you propose to do. If the agent asks what the land is for tell them you are moving into the area and want it for horses and it's easier to buy the land first, then the house near the land. You will, of course, have to put up with being mailed potential houses to buy from the agent. The farmer who is selling the land will feel cheated because he could have done what you have, except he didn't know how. Land is a funny thing, once you own a few acres it will become part of you and you part of it. Good hunting

BUYING

I suggest that you use a local solicitor who is well versed in buying and selling agricultural land in the area where the land is. The agent should be able to recommend someone.

PRIVATE TREATY.

This is where a house or land is offered for sale at a price which is open to offers. Even if an offer is accepted, it is not binding till contracts have been exchanged and is not yours till completion.

PUBLIC AUCTION.

This is where the piece of land is offered for sale to the public at a fixed location normally a pub or similar and bids can be made till the fall of the hammer after the highest bid. This is of course a simplification of the procedure. If you decide to bid on a piece of land at auction you will be required to get a legal pack from the auctioneer. This will cost you money and you will need to register and prove that you have the funds to complete the purchase in the allotted time (normally 28 days). There will also be a deposit to pay on the fall of the hammer. At this point you will have entered into a binding contract the equivalent of exchanging contracts in private treaty purchases. If for any reason you fail to complete, unless by mutual agreement with the vendor, you will lose you deposit and the land will be put up for auction at a later date. Any shortfall in the selling price, will have to be made up by you plus the entire expense of the auction. So do not bid if you cannot complete. A word of warning do not get "auction fever" decide on your top price and stick to it come what may. Remember it may adjoin a farm and the farmer wants it at any price. Don't compete, leave it

SEALED BIDS and TENDERS.

This usually happens when either a lot of interest has been shown in the land when it has been for sale by private treaty or when land is selling well in an area and the agent is trying to find people to pay a little more money than it is worth. It makes you wonder how these agents can value a piece of land and advertise it at a price, Then because it was too low he then changes tack and asks for offers above this price. Would you want this person selling your land. I wouldn't! Anyway back to the plot, if you wish to get involved in this dubious practice, you will need to contact the agent and get a tender form, you will need to prove you have the money to complete the purchase. Now as the agent wants to get as many people involved in his little scheme as possible, he will set a closing date (at least 6 to 8 weeks away) so you have to sit and wait to see how you have got on. If you have won, you can now start the legal work for the purchase. If you haven't, what a waste of time, but don't despair you will probably get another go. The winner may drop out when they discover they have offered too much money for the land. The land is then put back on the market and the pantomime starts again, why do people use these agents!

FORMAL TENDER

This is the same procedure as sealed bids but the bid you put in is binding. You have in fact exchanged contracts. No backing out and, as with an auction, a fixed completion date. This method of sale is rarely used these days, even though it is better than the informal tender method.

Good luck in your search for your land. Remember, buy the land you want to live on, not the piece you think you should. You will be living there a long time. Your search may take a long time but don't get disheartened there is a piece of land out there with your name on it and if you are meant to have it you will.

INSURANCE

Once you have bought your land, you will need to think about insurance. You will need to use a specialist broker who should be able to provide cover for all your farming needs, for the mobile home, the barn and to cover the building works, and farm machinery. Make sure to give all the information, particularly if machinery is being hired. These brokers are to be found in magazines about smallholdings.

PLANNING

The planning method we are going to use to live on the land is to build a barn and, while we are building it, we are allowed to live in a mobile home on site. As long as you are involved in its construction, it does not say you have to work on it full time and the only limit you have is that you must complete the building in 5 years. The barn will be built under permitted development which is deemed planning permission. Which is granted by an act of parliament not by the local Council and is perpetual; we do need to inform the council of our intention to build.

The way that the planning system works is quite simple Parliament makes and passes the planning acts into law. The government then issues a series of policy documents and statements which the local authorities have to implement to make the planning laws work. The planning act we will be using The Town and Country Planning (General Permitted Development) Order 1995 part 6 class A and the policy documents which cover this section are Planning policy statement 7. I will deal with the building first and cover the planning for the dwelling in a later section of the book. The criteria which allows you to build the building are dealt with as they are listed in the act. I will add notes if necessary so that you will understand what the planning officer will be looking for.

1) Must be a minimum of 5 hectares of land, this doesn't need to be a field it can be a mixture of pasture or arable and woodland. It must be agricultural land and to be used for an agricultural trade or business and the building must be reasonably necessary for the business.

2) The building must not be within 25 metres of the metalled part of a trunk or classified road. This just means from the edge of a road, I can only conjecture that this margin is left so that if at any time the road needs to be widened, the building would not have to be demolished.

3) The building must be a barn designed for agriculture the courts have held that the physical appearance and layout of the building and not its function are what define it. I have seen instances in the past of agricultural buildings built under permitted

development that looked remarkably like houses and long legal battles ensued to get them demolished .

4) The building or structure must not exceed 12 metres in height or 3 metres within 3 kilometres of the perimeter of an aerodrome. The reason is obvious for this, we don't want to be hit by any aeroplanes. The building that you will require will need to be more than 3 metres high, so buying a piece of land within 3 kilometres is really not an option.

5) The ground area of any works or structure must not exceed 465 square metres.

6) No existing building (if one exists) has been built in the last 2 years under permitted development rights, within 90 metres of the proposed building. This means that there is a time delay in permitted development, you can build a new building every 2 years or more than one if they are at least 90 metres apart. This can be quite useful if you want to build a barn and a polytunnel both under permitted development.

7) The building you propose will not house animals, It will be for storage of farm machinery and animal feed.

8) You will be using the existing access on to the land.

9) You will not remove any soil from the site, any surplus soil resulting from the engineering works will be stored on site for future use, the storage area must not exceed 0.5 hectare within the unit.

10) The barn once constructed must be used for agricultural purposes for a period of 10 years or it will have to be demolished unless a planning consent is given for another use.

11) No business rates are payable on the building or the land.

12) The building must be completed in 5 years from the receipt of the 28 day notice, the notice is dealt later in the chapter.

Permitted development is the same as having been granted outline planning permission by the government. The only input from the local planning authority are the reserved matters (these are the details of the development), if the conditions for permitted development part 6 rights are met. The local authority cannot consider if the development can go ahead, the right is already conferred. Local planning authorities may only concern themselves with;

1) The siting, design and external appearance of a propose new agricultural building and its relationship with its surroundings;
2) The siting and means of construction of roads.
3) The siting of those waste deposits which would exceed 0.5 hectare.

A site meeting with the planning officer will resolve these issues, and once agreed variations to the drawing should be confirmed by the planning officer in writing.
It would be a good idea not to mention to the planning officer what you are proposing to do on the land.

It would be a good idea before you meet the planning officer to contact a
Few suppliers of buildings to get prices for sizes of buildings and samples of the cladding and roof covering material so that you can show the colours that are available. The supplier will have done this before and, if he is local, he will know the colour the council will pick.

You will only be allowed to build a building of modern construction, normally of steel frame construction, plastic coated roofing and cladding sheets and concrete block walls. You will not be allowed to build say an oak framed or even brick built barn .The reason for this is about 5 years after its construction you could apply to convert it into a dwelling you would probably get planning permission.

Before you can start you will probably have to inform the local authority of your proposal to build a barn. This is done on a agricultural prior notice (28 day notice) to determine as to whether their prior approval will be required for certain details (General Permitted Development Order, Part 6, A2(2) and (3). The local planning authority have 28 days for initial consideration of the proposed development. Within this period they may decide whether or not it is necessary for them to give their prior approval to these details of the development involving new agricultural buildings

The determination procedure firstly establishes that part 6 applies and if the criteria is met, that a planning application is not required. There is no scope to extend the 28 day determination procedure, nor should the second the discretionary second stage concerning the approval of certain details be triggered for irrelevant reasons.

Provided all the permitted development requirements are met, the principle of whether the development should be permitted is not for consideration, and only in cases where the local planning authority considers that a specific proposal is likely to have a significant impact on its surroundings would the secretary of state consider it necessary for the authority to require the formal submission of details for approval.

The 28 day determination period runs from the date of the written description of the proposed development by the local planning authority. If the local planning authority gives notice that prior approval is required they will have the normal 8 week period from the receipt of the submitted detail to issue their decision, or such longer period as may be agreed in writing.

Requests for more time from consultees should not be used as a reason for requiring the submission of details.

The planning authority should acknowledge the receipt of the 28 day notice in writing, giving the date of receipt; I would check that they had received it by phone and that it contains all of the information that is required. The planning authority should inform you

If they do not require the submission of details at the earliest opportunity not wait till the end of the 28 days. It would be beneficial to include not only the agricultural prior notice duly completed and the fee, perhaps a sketch showing the proposed elevation of the building an advertising pamphlet from the building suppliers showing selection of cladding Colours and the design of the building. Any variations to your original application do not need to be resubmitted formally, any agreement made between you and the planning officer regarding any changes he should confirm in writing so that you can proceed If the local planning authority needs to carry out a technical appraisal or call in consultants, where this is necessary this should be done in the 28 day period and not simply call for detail on a precautionary basis. If you do not hear from the planning authority, other than the receipt letter for your application after 28 days you may proceed with the works.

The siting of a new agricultural building can have a considerable impact on the landscape; developments should be assimilated into the landscape without compromising the function of the building. Siting the building on the skyline should be avoided if possible, to reduce their visual impact buildings should be blended into the landscape or, on sloping sites, set into the slope if that can be achieved without disproportionate cost. Resiting of the building on another part of the land may be a solution, if this can be achieved without undue operational difficulties. Where constructional problems emerge after proposals have been notified or approved, authorities will need to take a flexible approach to requests for approval of departures from the original proposals. The siting of the new building adjacent (not too close) to existing woodland may help assimilate it into the landscape. Elsewhere judicious tree planting and landscaping may help, the idea is not to hide the building, but to soften it in the landscape. Any new planting should reflect the vegetation type already existing in the locality, or be part of a woodland grant scheme application.

Roof overhang reduces apparent scale, as does the use of different materials for roof and walls. Well designed features such as rainwater downpipes and gutters, ventilators, eaves and gable overhang, emphasise the shape of a building. The colours chosen for the building should be compatible with the rural setting, not to camouflage the building. Careful choice of colour reduces the apparent scale of a large agricultural building (if the roof of a building is coloured darker than the walls, the visual impact on its surroundings is reduced).

ARTICLE 4 DIRECTION

This is an order made by the Local Planning Authority and needs to be confirmed by the secretary of state, it is generally used in a conservation area to remove the permitted development rights in that area. This can also extend to environmentally sensitive areas like national parks or areas of outstanding beauty, and can be used to restrict or remove permitted development rights, it is unlikely that you will find a piece of land with restriction on it, even so be careful and check.

THE DWELLING

The mobile home is allowed on the land under permitted development and you are allowed to live in it whilst you are engaged in building the barn, this is covered at the back of the book and should be read in conjunction with the caravan act, also reproduced with it. I am sure as you have read through this book, every time you have seen the words "mobile home" your heart has sunk. Well, don't despair, the definition of "a mobile home is a caravan, a wooden structure which can be easily dismantled or other temporary accommodation". The selection of the mobile home is crucial. Your comfort in the coming years will depend on it. The level of comfort you achieve is as always dictated by your budget. Most people think of buying an ex holiday caravan to live in. This is a mistake as these caravans are built as holiday caravans to be lived in, during the summer not all the year round. They do not have any insulation and in the winter you will freeze. They are also over ventilated at floor level because propane gas is heavier than air. In the event of a leak, the gas goes to the floor and out of the vents. To be on the safe side, they fit loads of vents and what lets gas out, lets draughts in.

The best type of mobile to buy is one off a permanent site. These caravans are constructed to building regulation standards, with full insulation. That is why when they are new they are so expensive. They are sometimes available second hand and are usually twin units. The single unit seems to have been replaced almost totally on permanent sites. The twin units are much more spacious, some with 3 or 4 bedrooms and good size reception rooms. A log cabin can also fall into this category and there are some manufacturers who make units for this purpose. A log cabin is almost a house, with all the comfort and amenities. Preferably buy one from a cold country, not from timber grown in the United Kingdom. Cold grown timber is slow growing, the rings are tighter together it is denser and less likely to twist. Its just better wood. When you decide on which type of mobile home you would like, you will need to order it. If it is new, there will be quite a lead in time to delivery. You will also need to know the size and position of the concrete slab it is going to sit on and its position on the land. When the mobile home is habitable, apply to have it council tax rated

The reason that we can live on the land in a mobile home is covered by The Town and Country Planning (General Permitted Development) Order 1995 Part 5 Caravan Sites and Caravan sites Control of Development Sites Act 1960 (c62). First schedule, paragraph 9 Building and Engineering Sites. Between the two acts of Parliament they state that no planning permission or site licence is required, for you to live in a mobile whilst you are engaged in the construction of the building. I have included a copy of the "Measor principle" to illustrate the true planning situation with regards to mobile homes on land. As you will see the situation is that you can have a mobile home on your land. According to this judgement and subsequent other judgements, the Local Authority is powerless to stop you. If you are confronted with a situation where the council tries to remove the mobile home under the town and country planning acts, they cannot. Reproduce the principle and present it to the council.

THE MEASOR PRINCIPLE

Planning permission is not required for a mobile home in the United Kingdom, planning permission is only required for "Development" of land. The definition of development is "the carrying out of building, engineering, mining or other operation in, over or under land or the making of any material change in the use of any building or other land".
The term "Building" and "Building operations" have both been defined by case law.
The term "Building" can include many structures and erections which might not ordinarily be regarded as a" Building" but the courts have decided that if something falls within the definition of a "caravan" it cannot also be a "Building". The definitions are mutually exclusive.
The legal definition of a " Caravan " is any structure designed or adapted for human habitation which is capable of being moved from one place to another whether by being towed or by being transported on a motor vehicle or trailer.
The definition varies in "planning policy statement 7" which is binding on planners "Caravan, a wooden structure which can be easily dismantled, or other temporary accommodation".

Prior to 1998, there was some debate in cases as to whether mobile homes were or were not buildings for the purposes of planning. The relevant Considerations are the factors of permanence and attachment. The matter was settled in 1998 in the case of Measor v. Secretary of state for the environment, transport and the regions. The case was decided in the Queens Bench Division of the High Court and is thus binding on planning inspectors.

In this case, the judge stated "in my judgement, it would conflict with the purpose of the act and common sense to treat mobile caravans as "Buildings" as of right".

A mobile home could never be a building for the purpose of the Town and country Planning act 1990. It seems to me that, by reference to the definitions in the caravan site and control of development act 1960 and the 1968 caravan sites act, it is clear that in the present case caravans lacked that degree of permanence and attachment to constitute buildings.

Thus in 1998 whilst the courts were anxious not to state that a mobile home or caravan could never be a "Building," in general they are not. The "Measor Principle" as it is known, has been restated and approved in a number of cases since 1998. As recently as 19th June 2002 Mr Justice Forbes in the case of Massingham v secretary of state for local government and the regions stated "the judgement in the Measor case restates the approach to the definition of a building for the purpose of development control which has been well settled in previous cases including Barvis limited v secretary of state. Acknowledging it would be wrong to say that a mobile home could never be a building. "In this case I am in no doubt that the mobile home fails the test of permanence and attachment established by the courts".

The facts of the Massingham case are worthy of note, the mobile home concerned was being used as a second home. It was supported by concrete slabs, although the wheels of the home remained, the axles were supported on blocks, and as Forbes j pointed out elsewhere in the judgement the fact of connection to mains services is not material in determining whether a caravan is a building.

It is now quite settled law that a mobile home would fall within the definition of a caravan and thus outside the definition of development.

CONSTRUCTION

The planning for the site layout should be thought about, You will need to agree a position for the building, trying to make it fit into the landscape as much as possible, you are not going to hide a huge tin shed by positioning alone, the best that can be achieved will be to put it up against trees, or plant trees round it when built, to soften it in the landscape. It will, of course, need to be at least 25 metres from the road. Try not to be too far onto the land, as this lengthens the access road which can be costly. Try to position the access road so that the electric overhead cables do not cross it, even if this means diverting the road. You can recognise the low power cables as they are normally fixed to poles not pylons and are quite thick not to be confused with telephone cables. This is where your power will be coming from so if you can see them from the land, this is the direction they will be coming from. If you cannot see them from the land then it's going to cost a lot of money but you will be able to tell them where you want the cables run within reason.

Do not try to move your access, as this will require planning permission involving both the local council and the county council. If you plan to have a touring caravan site, try to position the building with one end facing the area you intend to use as your caravan park. It can then be fitted out as toilets for the caravans. Next to the toilets you will need to construct a room, probably 10 metres by 6 metres, to be used as an animal feed store or an area to sell farm produce, hanging turkeys at Christmas, a general room with concrete floor, block walls all to be tiled, about 3 metres high with a plastered ceiling. The power for the holding will come in this end of the building so a small area will be needed for the meter and the consumer unit (distribution board). All power to the mobile home and the caravan park will come from this board. If you propose to use the roof area of the barn for solar panels, ideally the roof would need to face south. The barn doors should never face the road for security reasons. When the doors are open, any one passing can see what you have in it and it is too much trouble to keep closing them when you go in or out. The position of the septic tank will need to be sorted out so it can take the sewage from the toilets in the barn and the mobile home. You do not want too long a distance from the discharge points to the septic tank as the falls on the pipes get greater and the tank gets deeper and, without very sloping ground, the discharge from the tank will be impossible.

THE EXCAVATIONS

One of the first things you must do when taking possession of the land is erect a framework of some kind, adjacent to the road so that you can affix large signs to it. Your first sign will be "free tip for clean brick hardcore" and it must be large enough for passing motorists to see easily. This is your free advertising which will save you a fortune, anything and everything you produce or any activity you have going on should be advertised on the board. If people don't know what you have got, they can't buy it. Hardcore you cannot have too much. When you dig out the top soil, the only thing between you and the mud is the hardcore. Hardcore is a by product of the building industry. Its use is limited, in fact I cannot think of any except perhaps temporary roads. Most hardcore ends up in skips it would be a good idea to phone the local skip hire companies who normally would have to pay to tip it, and offer them a free tip. They may offer to sell it to you. No one buys hardcore unless they are in the middle of no where. Make it plain you can take an awful lot of the stuff but you only want clean brick hardcore no concrete, no bits of wood, rubbish etc. Before you start any excavation work, you should check with the utility companies. These are the companies who own the infrastructure, not the supply companies, to see if any services run under your land. You will need to send them a map of the land with an ordnance survey reference number and ask for a reply in writing confirming there are no services or, if there are, their location and where they run. You will need a copy of this letter if you hit something, as this can be expensive, better to be safe than sorry. Any services that you may come across in the ground should be at least 600 millimetres deep, though in practice some could be just below the surface. In the case of an electric cable, clay tiles used to be put over the cable with "danger" on them, a practice that has now ceased. All services are now colour coded: blue for water yellow for gas and for electric mains cables (if they pass through a duct) it must be black as is the colour of the cable.

Take pictures of the existing access on to the land from all angles and sides and keep them as a record just in case the council try to accuse you of changing the position or size of the access at a later date. For security purposes you will need a fence along the road side of the land. Get a contractor to do this. If you watch him do it, you will know how to do it yourself on the rest of the land and what equipment to buy to do the job. It will also get the front secure quickly. It is good practice to move the gate position back on to the land about 20 metres. This allows vehicles to enter the site without opening the gate; it is far safer for the road users as it causes no obstruction to the highway. The same when the site is being left, the vehicle can pass through the gate then it can be closed before the vehicle fully leaves the site. The reason it is 20 metres is to allow for articulated Lorries and cars pulling caravans. The access road needs to be at least 6 metres wide. Mark the road position out using pegs and a line, get a bucket of cement powder. Using your hands, scoop out the powder and using the line as a guide, allow the powder to run out of your hand to form a solid line. Only touch cement powder with gloves, preferably rubber or something impervious. Marking out the road should only take place immediately before the digger is ready to dig or it will get washed or blown away. Remember to remove the line and pegs before the dig begins. Only remove the topsoil not the subsoil. top soil should be banked on one side of the road. If the road runs along a boundary or both sides (if in the open), you will also need to use some of the soil that you excavate from the apron in front of the barn. Any top soil you dig out that you have no immediate use for needs to be stored. When the barn is completed there will be areas round the outside that will need to be top soiled, it can always be used to level out uneven places, never remove it from the land.

    I am jumping the gun a bit here, as you will need to select the type of excavator. You will need the choice will be dictated by the weather. If the ground is dry and the forecast, when you are going to do the dig is dry, then a JCB 180 degree rubber tyre digging machine will do the job. This the cheapest digging machine to hire lowest hourly rate and no transport costs as it travels on the road under its own steam. If the ground is wet and likely to stay wet, then the answer will need to be a 360 degree machine on tracks 12 tonne or above. This will need to be transported on to the site which can be expensive particularly over any distance. It may be possible to hire either of these machines locally from farmers. Most farms have a digger about the place, it would be cheaper to use them as they don't have the overheads of a plant hire company and use a tractor and trailer to deliver the machine. You are also going to need something to carry the soil about the site. You may wish to hire a dumper, if so, get a very large one. You will need to carry a lot of soil or the digger will be under utilised. The other alternative is to buy a tractor and tipping trailer. These can be purchased from agricultural machinery auctions and farm sales, which occur regularly throughout the year in all areas. The advantage of this is you save the hire money and put it towards the tractor you were going to buy anyway. Buying a tractor is relatively simple. Stay away from the small tractors, only buy a tractor which accepts normal size farm equipment, which can be bought at any farm sale or borrowed from neighbours. Find a tractor that is not packed with electronics, something that can be fixed without the need for computer diagnosis. Your ultimate aim is to service and repair your own tractor. Firstly, you will need someone else to do it so you can watch. There is always a local a part time tractor mechanic you can use. Farmer's sons are the best and work for sensible wages. Whatever your decision regarding purchase or hire, the machine will need to be parked away from the land over night until you are living there, you will need a farmyard that is pretty near and a friendly farmer who, for a small fee, may let you park in his barn.

  The other alternative is to hire the machine and dumper from the same company with a driver. If the situation is described to the hire company, they should have a machine with vandal guards fitted, which is virtually thief proof. The digger will then put its bucket in the skip of the dumper and stop this being stolen. The difference between the two types of digger, besides the output, is the way they work. If the ground is wet the JCB, being on tyres, churns up the ground and, as it has quite a short reach, it needs to keep repositioning its self to dig. The 360 degree machine has a much longer reach and can be positioned to carry out a lot more work without moving and if it does need to move (being on tracks) it makes less mess. When moving the soil away from the excavated area, always stay on the areas to be excavated. It is far easier to do this than reinstate churned up ground, the dumper can make a great deal of mess and can compact the soil, which doesn't help the ground to drain. If you are running your own tractor and trailer you will need diesel and a bowser to store it in. It will be better to hire one in to begin with. A minimum size of 500 litres this will be on wheels so that a road tanker can fill it up at the road side as the lorry will not be able to travel over the excavated areas without getting stuck. The bowser can then be towed to any part of the site by the tractor for refuelling purposes. If you have the machines on hire with a driver the hire company supplies the diesel.

  You will need a programme of work; what is to be excavated and in what order. A lot will depend on the supply of hardcore. You will need a solid road to the over site of the building, without this road you will not be able to get a concrete lorry onto the site to put in the pads for the barn, and certainly not erect it, as this operation needs a lorry mounted crane. It is cheaper to carry out all of the excavation work at the same time; the access way, the over site for the barn, the apron in front of the barn, the mobile home base and the hole for the septic tank. No trenches for service runs as these will be done with a trenching machine and a mini digger at a later time. You will need a set of drawings from the building manufacturer to get the position of the pads and the position of the bolts set into the concrete for the base plates to be bolted down on to. The pads are foundations for the steel frame of the barn. They are concrete blocks that we cast in the ground with bolts protruding out of the top which the base plates of columns of the building bolt to.

THE BUILDING

If you don't want to use a surveyor to set out the site you will need to get conversant with some type of level, something that will measure over a distance of at least 50 metres in day light I say this, as some of the levelling systems being lasers are hard to read over distances in sun light. Once conversant with a system, whether bought or hired, and a long tape at least 50 metres in length, you are ready to set out the position and levels for the building. To form the pads we will use a wooden shutter, a box of the right dimensions open at each end. You will need to make up all of the shuttering boxes before hand as they will all be installed at the same time by the digger. They will be constructed out of 18 millimetre ply wood. There is a grade called shuttering ply use this it is a lot cheaper. We then dig the hole for the shutter and lower it in to the right depth. I suggest that you set the pad height 250 millimetres above the existing ground level at the highest point of the surrounding ground. Then you can adjust the shutter height to the right level and fix it into position by nailing timber across the sides of the shutter, with the timber resting on the ground then back fill with soil behind the shutter with loose earth trying not to move the shutter too much. This is not too critical so long as the bolts positions will end on the pad somewhere near the middle. This operation is then repeated for all the pads in turn. Each one is levelled back to the first one not from the last one you fitted because if you make a mistake with one you will carry it on so all of them will be wrong. Always check the triangulation between the corners and the measurements between the pads both ways. When all the pad boxes are in you can look at the bolts and their positions. We take a piece of 18 millimetre ply this needs to be 50 millimetres larger than the widest position of the bolts mark out on the board. The position of the bolt (from the drawing) the centres and drill the holes to take the bolts. Now put the bolts through the timber with a nut on the bolt either side of the timber. This will allow the bolts to be adjusted up and down to the right position. The bolt heads, of course, are at the bottom and will end up in the concrete to the right depth. Once adjusted this piece of ply is then fixed on to the top of our shutter box by nailing timbers across the ply, the ends nailing onto the top of the box. The bolts can be adjusted for height and tightened to clamp them on to the board. Do not put too much timber across the top of the shuttering box as concrete from a lorry mixer needs to discharge by chute into it. You need to check and double check all of the measurements against the drawings to make sure they are right and check all the levels to make sure they are right. In fact it is safer, if possible, to get some one else to check them. As the carpenter says "measure twice, cut once".

Before the erection of the barn can take place by the company that is supplying it, unless it is exceptionally dry, you will need to have hardcore on the access road and the area of the barn and the apron to allow access for the lorry and crane. The apron needs to be large enough to turn a lorry around. Once you are ready the frame can be erected and the roof put on including guttering and the side cladding, leaving a 3 metre gap down to the ground. This is where the block work goes. While this is happening you can get on with shuttering and concreting the slab for the mobile home base and the machine can dig out the hole for the septic tank. The details and size you can get from the manufacturer of the system you have to decided install. The installation will need to be passed by building control, as will the pipe work from the toilets in the barn and the mobile home. You can now off hire the large digging machine and the dumper

Once the barn is erected and clad you will need to start digging the footings for the block work walls. For this we will need a mini digger, it is far more manoeuvrable than the large digger and causes less mess. They are cheap to hire and you can drive it yourself. The footings will run between the pads at the level of the ground. These don't need shuttering as the ground will be the shutter. The size of the footings will be 450 millimetres wide by 600 millimetres deep. You will be using 100 millimetre concrete blocks. To get the position of the centre of the footing, hang a plumb line down from the face of the cladding then measure the thickness of the cladding and add 75 millimetres from the plumb line. Measure inwards the size of the cladding plus 75 millimetre and drive in a peg. Do this at each end of each bay between the columns, string a line and mark off with cement as we did for the road, then dig. When you have dug out sufficient trenches to take 5 or 6 cubic metres, you can order the ready mix concrete. Do the area at the end where the toilets are going first. You will also need a trench across the building to carry the back wall of the room and the toilet. Once this is all concreted the first course of block work can be laid, I suggest that you get a bricklayer to build the room and lay the concrete slab. This will speed up the whole process. Firstly you need the room so that the power supply can be connected. Secondly you can watch the bricklayer set out and lay the blocks. I am a great believer in monkey see, monkey do.

The room will comprise the toilets gents and ladies. Make these large about 3 metres by 4 metres, big enough to take 2 toilets, 2 shower cubicles and 2 wash hand basins in each. When you are setting out the end of the building you will require 1 door opening from the room to outside, 1 to inside, and a door each for the toilets from the outside.

You will need some opening windows for the toilets, it will be better to get these first. A cheap source of new windows is to phone round double glazing companies to see what they have got. They sometimes make windows the wrong size or the person cancels the order, as there is not a lot they can do with them, you should get them cheap. You can build in any size of frame, with no existing size constraints. Always use concrete lintels on the barn as they are a fraction of the price of catnic lintels. It would be a good idea to lay one or more courses of block work to get a height for the slab. You can use the blocks as a shutter and a surface to tamp along. During the laying of the slab you will need to put sheet reinforcement into the top of the slab to give it strength. Once the slab is laid it needs to be as smooth as possible as it will eventually be tiled. Once the slab has gone off (this will take quite a few days), it may have set but it is not really hard. The longer you stay off it, the better. The walls can be built up to the cladding it should pass up the back of the cladding. It should over lap by about 1 block. The rest of the walls without footings can be built off the slab. Once all the walls are up and level a plate can be fixed to the top of the block work and screwed in place using plate straps. They are about a metre long and bent at a right angle at the end. They have predrilled holes so they can be screwed to the plate and screwed to the wall. These will need to go in every 2 metres or so. The joists can now go in for the ceiling span them the shortest distance between walls, this keeps the cost down and you may be able to pick some up second hand from a demolition contractor. Always find out the cost of new ones, then phone round see if you can get them cheaper. The plans you have for this loft area, dictate the size of the joists you use. If you are going to board it and use it for storage or just leave it, if its for storage use 225 millimetre by 70 millimetre or near it and board it with 15 millimetre shuttering ply. If not to be used as storage, 100 x 50 to secure the plaster board to.

Once the room is built it can be plastered, the block walls will need to be rendered on the inside with sand and cement. This strengthens them and gives a good surface to tile on. Plasterers do not like doing this as its hard work and come up with all sorts of excuses, not to do it. If you use plaster, the damp will attack it. If you want a plaster finish anywhere, the plasterer should be able to polish the sand and cement and skim it with multi finish, the ceiling can be just skimmed. The floors and walls can now be tiled, all of the electrical wiring and the pipe work will be surface mounted. It is easier for maintenance and any layout changes you may wish to make in the future. The wall where the distribution board is to go will need to be fitted with a full sheet of 18 millimetre ply. This will need to be screwed and stuck to the wall with builders adhesive.

The floor of the barn can now be tackled. You have a hardcore surface which will need blinding in to form a compacted surface. We can do this in two ways. Method 1: using M.O.T. type 1. This material is 18 millimetre chips, down to dust, can be levelled easily and compacts with a Wacker plate. Method 2: use asphalt grindings. These are the result of road grinding which happens to most roads before they put a new top on it. If you phone around large asphalt supply companies they should be able to give you the names of a few haulage contractors who carry it away for them. The grindings are not so easy to work with but do compact with a wacker plate. The difference between the two is price. The type 1 you will pay for the product and the haulage, the grindings you will pay for the haulage and a nominal sum for the product. Always find the closest place for both products to reduce costs and always have it in 20 tonne loads. It may be possible if any sizeable contracts for road grinding are going on in the area to take it as and when it is produced. As you have the place to store it you could do a nice deal. The access road and the barn apron will also need to be done.

If you decided to concrete the barn floor all over it would be a sizeable investment and would probably cost almost as much as the frame cost. Unless you have a specific need I suggest you use the methods above.

When you contact the electrician get him to calculate the power requirement of the site so that you can tell the power supply company the size of supply you will need. Make an allowance for expansion of the site, particularly if you are having a touring caravan site and would like to expand it in the future. It is better to do this to start with than come back in a couple of years and pay out another load of money to take out what you have put in and put in a larger supply.

THE SERVICES

You may have no choice but to have a bore hole for water. This is not as bad as it sounds as it would pay for itself in a few years. Bear in mind that you are going to need a pump to lift the water. Some pump set ups only pump on demand, perhaps when a tap is turned on. This system is fine but you have no back up. If the pump fails, you have no water: a dire situation if you have animals. An alternative is to have a storage tank in the roof of the barn, if the pump then fails, you have some storage. The choice is yours, you will need to decide when the building is being fabricated at the factory so that a frame can be fitted to carry the tank and the tank will need to be on site so it can be lifted into position by the crane before the roof goes on.

The services can now be put in using a trenching machine. This cuts a trench 150 millimetres wide and can move around under its own steam. The reason for using it is that it does not make a mess and the reinstatement is minimal. You will need to run trenches for the sewers from the toilets in the barn and also from the sluice, which you will need to build outside the toilets, to the septic tank. There will need to be a manhole where any pipes join the main run and every 12 metres, so try to keep the runs short. All pipes will be plastic and will be surrounded by pea shingle to protect them. Do not cover up the pipes till the building inspector has seen them. There will need to be quite a lot of water main pipes. Never put joints in the ground, always do the long runs first where you have branches to feed like, taps. Try to bring the pipe up out of the ground, put on the fitting then a new piece of pipe to the next joint, so all the joints are out of the ground. The water will need to come from the borehole or the main supply in the road to the barn, and get redistributed from there to the mobile home and the caravan site pitches. The electric supplies will all come from the barn to the borehole pump, to the mobile home and then to the caravan pitches. Just a note all water pipes and cables must be at least 600 millimetres below the ground to prevent freezing in the case of the water pipes, and safety in the case electric cables.

THE BUSINESS

To get planning permission for a house we must pass the three tests: functional need, viability and sustainability of our business enterprise. Most people fail the tests because they want to do just one thing, like just keeping goats. This would pass the functional need test on care and welfare, but would not be viable as you could not keep enough goats to pay you a wage unless a substantial investment was made to add value to the milk, like producing cheese or ice cream. It would need to be on such a scale that the smallholding would resemble a factory. The functional need test can be met with goats and you can derive some of your income from them, but you must do other things to generate income up to a level that passes the viability test. You could have a poly tunnel and grow all sorts of things, you could also keep chickens. I am sure you get the idea. The only consideration is time management. Do not engage in activities that take up lots of time for little return, each enterprise will need to pay for its self and give a good return. There is no point running your self ragged, it would not be fair on the animals or you.

Good husbandry of animals is not just about feeding them and cleaning them out, it is also about taking time to watch them. This is how you find out if anything is wrong. It should be possible to build an agricultural business that should meet the tests; you will require other skills to succeed. Besides being a producer of farm products you will also need to retail them to make any money and add value if possible to any product you produce.

The difference between a farm and a smallholding is not just size, it is where the produce goes. Farmers sell into the wholesale market and smallholders must retail. That is why I keep banging on about the land being in the right place, so that you can advertise for free. There is no point in producing anything if you can't sell it and you won't sell it if no one knows it's for sale. The big thing in the countryside is tourism and this is a must to tap into. It can be anything from a pick your own enterprise, to a touring caravan site. Always remember if you can get people onto the smallholding, you should be able to sell them something.

A brief word about farm machinery if you are using a normal size tractor as the motive force around the smallholding, then the layout should reflect this. Do not create tight corners or inaccessible places. Do not create situations that require specialist small equipment.

I have put a list of businesses in to give some idea of the enterprises you can carry out on the smallholding. Its not exhaustive and about most of them I have very little experience, remember to pick a main enterprise that is animal based and fulfils the functional need requirement. All the rest you can add on as you like, try to pick seasonal things so you have income all the year round.

In any start up situation, the first thing that everybody tells you is, "write a business plan." This is going to be difficult. If you wrote a business plan for every enterprise you are going to undertake on the smallholding you would probably be writing full time. Most of the enterprises you will be involved in will involve a small outlay financially, and with research, should be successful. If you tailor what you wish to produce to the market that you have, there should be no reason why you will not sell them. In other words don't try to sell Noordman Christmas trees at £30 in a poor area, or cheap trees in a wealthy area. Turkeys fresh from the farm always sell at a premium to the market price, if you are in doubt what that price is then phone a local butcher. Try to find out where people buy things in the local area and only go into things that there is not too much competition. Its not rocket science, buy as cheap as you can and sell for the best price you can get. Don't go into enterprises in a big way in the first year, it's the only way to judge true demand. Your second year you should have a better idea of demand so you can go for it and make some real money. You will make mistakes, just try to make them small ones.

I have included a list of businesses starting with the ones that that fulfil the functional need test and should pass the viability test, if some other enterprises listed separately are also operated along side the main animal business this should be a formula for success.

I am not an expert on any of these businesses, so I have given a brief out line of what the business could be both good and bad. The decision is yours you must base that decision on your research of the local markets. When you decide on an animal enterprise (as you have to succeed in getting planning permission for a house) then learn as much as you can about those animals get books, go on courses, visit dealers, ask questions, this will make life so much easier when you get the animals.

THE TESTS

FUNCTIONAL NEED

This is the reason you need to live on the land for the health, care and welfare of your agricultural animals. Security can also be a reason but is not sufficient on its own. The definition of an agricultural animal in this instance is one that produces meat, fibre or milk, and also to pass the test needs to have a random breeding cycle. This means it does not have a birthing season like sheep which would not pass the test as you would only need to live on the land in the lambing season, not all the year round. Incidentally the sheep would fail on the viability test as not enough animals could be kept on a smallholding.

VIABILITY (FINANCIAL)

You will need to prove that the holding is viable. A set of accounts (or a copy will suffice), which has been submitted to the Inland Revenue. It should show that you earned the minimum wage for 1 year out of 3.

Prevailing legal minimum wage per hour x 40 hours a week x 50 weeks of the year this will give a figure of how much profit you will need to earn to pass the 1 in 3 part of the test. You need to earn as much money as possible as there is a relationship between your income and the size of the house. You will be allowed to build regardless of your personal wealth, on the minimum wage the house would be very small. Always remember the total income from the holding is the figure you will be putting in from all of the enterprises, not just the main animal business.

SUSTAINABILITY

You will need to pass the sustainability test to show that the business is likely to last in the long term. This is normally demonstrated by additional investment in buildings or stock, adding value to your produce or even diversifying further into other enterprises. It is difficult to prove sustainability however, if you have a thriving diversified business, who can argue with that.

MAIN ANIMAL ENTERPRISES

ALPACAS

These animals fulfil the functional need test as they are fibre producing, which makes them agricultural, and they give birth at any time of the year. The reason this is important is that they need care and attention at all times. Whereas with sheep, as an example you would be allowed to live on the land but only in the lambing season, and you must also have a permanent place to live elsewhere. The viability test can also be met, as these animals have a high stocking rate up to 6 per acre on really good land less of course on poorer pasture. The value of the fibre produced is quite high but the yield per animal is quite low, however the money is to be made from breeding and selling on the progeny, most of the sales are selling pregnant females for quite large amounts of money. This is a long term venture as the gestation period is over 11 months and the females cannot breed until they are 18 – 24 months old, there is no fast money in alpacas,

Another way of earning money is to hold courses in the care and welfare of the animals for people who also wish to keep them. To be viable you would need a herd size of about 30 Don't be put off by this as they are gentle animals easily handled, they will require moveable field shelters in the enclosures (planning permission is not required) the reason for this is the fibre on them is not waterproof and they need protection in wet conditions, They will also require access to a good clean water supply in each enclosure. Some breeders house the animals in winter, this will of course depend on climatic conditions, the Alpacas have adapted well to UK farming systems and would even be suitable for upland or hill farms, stocking levels stock levels would need to be adjusted according to the quality of the pasture.

Capital costs are quite high with pregnant females costing between £2000- £8000 and stud males £2000- £9000, all dependant on the quality and the lineage of the animals.

Annual running costs can be off set by the sale of the fibre, but the main income source is animal sales.

If you decide that you would like to keep alpacas then go on a short course to learn about them, details of these courses are in magazines about smallholdings or on the web.

Alpacas are low maintenance animals and would make a good main animal enterprise along with seasonal things to supplement your income.

FARM ATTRACTIONS / PETTING ZOO

This enterprise can fulfil the functional need test if the scale of the enterprise is large enough, and involves a large amount of animals and a lot of diversity of species, what would be required say pigs, goats, donkey, chickens, alpacas, in fact any type of farm animal plus pets, rabbits guinea pigs any dwarf animals are a big favourite, animals should always be kept in groups never singly. No licence is required so long as the animals that are kept are "animals that are normally kept in the U K" a licence may be required if birds of prey are kept or exotic animals like ostriches. You would need to carry out quite a lot of local research to make sure that there is not too much competition locally, you would need to spend significant amounts of money on advertising the current rate would be about 7000 visitors generated from every £1000 spent on advertising. There would also be additional costs for creating car parking, animal and human friendly fencing and washing facilities outside any enclosure where any animals can be touched.

The up side is you can generate lots of income from the visitors catering is a good source, use a catering trailer as it doesn't need planning permission. You can grow a maize maze, do tractor rides, grow pumpkins ready for Halloween, sell turkeys and Christmas trees. Even out of season put up notices so everybody knows what you do or will be doing in the future. These attractions would also fit in with a touring caravan park, the only downside to this enterprise besides the set up costs is the labour costs which can take as much as 50% of turnover so a large family would be handy. The second point if you are going to be open seasonally the animals require the same level of care and of course food in the winter as the summer. If you are planning to be open all the year round then you will need to put in hard standing areas outside the pens and an extensive network of footpaths. Other events can also be run at various times of the year Archery, Clay pigeon shoots, hot air ballooning and craft fairs.

Rare breeds could be a theme of an attraction farm and also enlarge the population of any animals whose numbers have been depleted over time as farming methods and modern needs have made them obsolete.

If you have woods on your land you could do paint ball days or you could hold shooting days, or if you have particularly undulating ground you could create a quad bike track, you will need to utilise your piece of land as much as you can to make money.

CHICKENS FOR EGGS

A free range egg enterprise would fulfil the functional need test, it could also pass the viability test, the problem is the size of the acreage of land required as the maximum stocking rate is 400 birds per acre under present legislation. To be viable you would require a flock of at least 10000 birds which would be at least 25 acres and the cost of the special building to house them would be very high, also the machinery to handle and pack so many eggs would add substantially to your overall investment. Your investment to enter this industry would be hundreds of thousands of pounds, and remember you would have to enter into a contract to supply all of your eggs to one packer who in turn will be supplying a supermarket chain which means you are at their mercy. If you like chickens keep some as an add on business not as a main enterprise, enough to feed yourself and perhaps a few to sell, see if sales build up if they do get more birds, always remember you must always retail never wholesale.

CHICKENS FOR THE TABLE

This as a main enterprise has exactly the same problems as chickens for eggs, this is factory farming on an enormous scale, you will not have the size of land problem but the investment required in buildings would be enormous, and you would be competing with producers from the far east and other low overhead countries. The supermarkets would of course be calling all the shots on money, do the same as the eggs establish a local market and grow it if possible you would make more money out of 50 birds selling them retail than you would earn out of 1000 birds in a broiler unit your margins in a broiler unit would be pennies per bird, and really did you go into smallholding to be a factory farmer.

WILD BOAR

These animals pass both the functional need and viability tests, it would be fair to warn you that these animals can be dangerous, and accordingly will require a licence from the local authority under the dangerous wild animal act 1979 and 1984. The cost of adequate fencing is very high as it needs to go down into the ground 0.5 m and also needs an electrified wire along the top and an electric fence in front, the cost would be approx £5 - £7 per metre. As the animals are kept outside they would also require shelters and arks. The labour requirement is quite low 1 man for every 50 animals. These animals stock at a rate of 10 sows per hectare the return would be about £ 500 per sow, the return would be dependant on the price of vegetables and of course the nutritional feed. The market for the meat is good it certainly sells to restaurants and is becoming more popular with people as people realise it has a lovely taste as normal bacon once did.

CALVES

Weaning calves will meet the test for functional need , the enterprise is buying in calves which are on milk and weaning them onto grass then selling them on to be fattened, it would be difficult to do this on a large scale and so on its own would fail the viability test but combined with other enterprises it would pass. I would suggest that before you decide to go into this enterprise you spend some time at an establishment that does this, you can offer your labour free in return for working with the animals and learning from the farmer.

PIGS

Once again this enterprise is factory farming entailing very large investment and very little reward and once again reliant on the supermarkets and their generosity, if you like pigs keep a few and sell them direct and of course eat them yourself.

OSTRICHES

These birds would will require a licence under "the dangerous wild animals act 1976" and have an annual safety and welfare inspection, the local authority oversees the licensing covered by the act there will be an annul charge for the licence. The fencing for the enclosures will need to be 1.7 metres high a fence or a stout hedge will do. Shelter from the rain will be required as the birds have little natural oil so some kind of building will be required, young birds when housed will require special heating arrangements. The stocking rate is 0.2 hectares per breeding trio for exercise and grazing, 2 hectares will be required for 25 young birds aged between 8 – 14 months old. Breeding trios will cost to buy £1200 - £1500 (young females who are proven layers and fertile males are worth £400 - £500 each) an incubator for the eggs will be required depending capacity will be £1250 - £4200. The returns can be quite high the meat wholesales £3 - £5 per kg and retail £8 - £13 per kg, the skin £35 - £120 per bird, day old chicks £10 - £15, 3 to 6 month chicks " £25 - £35, fertile eggs for incubation £5 - £10, infertile eggs for cooking and crafts £4 . There is a good market which at the moment is being met by imports the meat is popular as it is healthy being low in cholesterol and high in protein, best place for retailing farm gate, farmers markets and the internet.

UNSUITABLE MAIN ANIMAL ENTERPRISES

Below is a list of enterprises that I would consider unsuitable unless you want to go into factory farming, where the investment in controlled environment buildings and sophisticated equipment would be substantial. I am not sure that this is what smallholding is all about, the enterprises would be intensive and the produce sold wholesale. All of these enterprises could be carried out on a small scale to suit you, as add on business.

Ducks for eggs

Ducks for the table

Dairy goats

Dairy sheep

Angora goats

Cashmere goats

Angora rabbits

Rabbits for meat

Guinea fowl

Mushrooms

I do not think it would be viable to keep geese on a large scale as there is not the market for them, a few perhaps around Christmas time to sell along with the turkeys.

ADD ON ANIMAL BUSINESSES

BEE KEEPING

Keeping bees is a low labour input enterprise about 30 minutes per week per hive from mid April to August. The set up costs are quite reasonable the hives which can be purchased new or second hand should cost no more than £200 per hive perhaps less when second hand. Personal protective clothing including suit and veil £40 - £120, gloves £5 - £20 and ankle protectors £10, smokers £20 - £40, hive tools £10 transit net about £30 and the big one, honey extraction equipment £200 - £2500. One hive should produce between 40 – 80 pounds of honey which should retail at about £2.50 per 1 pound jar, honey extraction is taken twice a year remember if you retail honey you will have to comply with current legislation.

TURKEYS

A must for any smallholder turkeys and Christmas trees a winning combination at the right time of the year, your chance to earn some real money when little else is going on. The time to take delivery of your day old poults is the beginning of July. Which means you will need to order them a few months earlier to make sure your supplier has sufficient time to produce them. Correct feeding is essential and advice should be sought from your feed supplier, what you feed them will change as they grow. The amount of turkeys you rear in the first year will be a guess you can do all the research you like locally and it will still be a guess, be sensible do not get more than 30 birds this should be enough to test the water both in looking after and selling them. A farm bred fresh turkey should sell at a premium to supermarket birds. It is easy to find out the price per pound for fresh turkeys by ringing the local butcher or asking in the supermarket, you should easily better this price. Always try to keep the birds free range as grazing is good for them. They are susceptible to worms and must be wormed every 6 weeks to keep them healthy, a little cider vinegar in their drinking water changes the ph balance of their gut, which seems to make it less inviting for the worms. The birds will need to have access to their shelter at all times in the day time, and housed at night. The food and water should be kept inside the shelter, and there should be lots of feeders and plenty of drinking water. This will cut out any competition between the birds as these birds can be very spiteful to each other, sometimes for no reason. The birds will need watching as they do have a tendency to attack each other. Once blood is drawn then all the others will join in attacking the injured bird. The injured bird should be removed immediately and put on its own until it recovers, and only when its wounds are fully healed should it be put back with the others. If you are averse to killing then you should get some one booked in to kill them at Christmas. You should really do this before you order the birds, it will be no good having 30 turkeys ready for the table still running around on Christmas day. The birds should be starved for 24 hours before they are killed this cleans them out and makes evisceration easier and cleaner. Once the birds are killed they should be plucked immediately and hung for about 10 days before evisceration this tenderises the birds and gives them a better flavour more gamey. The birds will be hung in a cool dry place away from anything that can contaminate the carcass. This one of the uses for the room we have constructed at the end of the barn. The environmental health officer at the local council will need to be informed when the room is constructed. He can tell you what is required in the way of fitments and finishes to comply with current legislation. He can also tell you what is required with regard to the killing of the birds, and disposal of any of parts of the bird not intended for consumption.

WORMS FOR COMPOST. BAIT AND FEED.

Worms as a crop may seem strange however it can be a very lucrative enterprise if you like worms of course. They are used as bait for anglers, a high protein feed for poultry and fish. They are also used for composting large amounts of waste material. The worms are self propagating; they do however need feeding, watering and protection from predators. The management of the enterprise requires a certain level of skill and attention to detail in particular the management of the compost heap, moles and weed seeds can be a problem.

The land required for each breeding pit of 1000 sq metres is 0.5 of an acre, the start up costs for 1 pit can be from £6000 to £12500 this is for breeding stock, beds and equipment. The returns are very good and should be in the order of £450 to £600 per week for about 20 hours work, contracts are available for the worms, there is also additional income from the worm casts which can be sold to garden centres as a component of mixed compost, another emerging market for the worms is domestic compost heaps and worm bins. Listed below are companies that may offer worm farm contract opportunities and can be found on the internet.

Wonder Worms
Organic Resource Management
Eco Group Holdings

HORSES

HEAVY HORSES

Unfortunately horses are not generally regarded as agricultural animals so in most cases cannot fulfil the functional need test, I say in most cases as heavy horses can, be it depends for what purpose they are being kept. However it would be difficult to make a planning case to build a house just because you have a pair of heavy horses to pull a cart or plough. It may be possible to keep a number of breeding mares and open to the public as a heavy horse centre either on its own or part of a farm attraction, other than this I can not see a way to have horses and get planning. Horses are generally frowned upon by planners even when you get planning permission for a house you cannot ditch every thing agricultural and open a livery yard there will almost certainly be an agricultural tie on the holding. This planning restriction is not only on the type of people who can live there but also the type of enterprises that can be run on the holding without planning permission.

If you are a fan of heavy horses go to a heavy horse centre and see what they do and see if you would like to do it too.

PONY TREKING

To make any money out of pony trekking is difficult and is only viable if you have Accommodation to let to the trekkers as most of your income comes from the letting. The smallholding would need to be in a good geographical position near a national park, on a bridleway system basically good riding country. The trekking season is only in the summer but the horses eat all year round and taking care of them doesn't stop at the end of the season.

LIVERY

It may be possible to open a small livery yard once you have planning permission for your house this would involve a planning application however if you read planning policy statement 7 the government are encouraging equine pursuits in the countryside and the council are bound by this statement. The livery business would have to be run as part of you overall smallholding enterprises. It is doubtful you would succeed in getting planning permission to change the use of the smallholding to an all horse enterprise. You will have got planning permission for a house because of agricultural need.

GROWING

You may decide to grow various plants as a way of making money. It would be possible to have a polytunnel if you wish this could be covered by permitted development. As a polytunnel now requires planning permission to erect . It would of course need to be at least 90 metres away from any other structure built under permitted development, it would just require a 28 day notice the same as the barn. The type of plants you grow will depend on the soil type of the holding, and also climatic conditions and altitude, what you can sell locally would probably be the overriding factor in what you grow.

Bedding Plants
Herbs
Lavender
Peppers
Tree nursery

PICK YOUR OWN

Pick your own with soft fruit can be a profitable venture, one of the drawbacks with pick your own is that the customers must come to you and depending on the size of the operation you propose to run quite a lot of customers would be required, with other produce you can if necessary take it to the customers by way of a farmers market. This makes the enterprise particularly vulnerable when you have short season crops like strawberries unless you can get the customers in, you could end up with a lot of fruit rotting in the fields.

CHRISTMAS TREES.

If your smallholding is in the right place then retailing Christmas trees is a must and if your selling turkeys as well it is the ultimate combination, the mark up on the trees is very good you should get over 100%.Christmas can be sourced from wholesalers or direct from growers it would be a good idea not to buy hundreds of trees and then get stuck with them see if you can do a deal to have the minimum amount that the seller delivers and see if you can get top up orders as you run up to Christmas in the event of you running out. As you get more experienced and would require larger numbers of trees they can be bought at auction, these auctions are normally advertised the Farmers Guardian news paper. Wreaths can also be made and sold with the trees; these are made from the trimmed branches of the trees or from the seller of the trees normally for nothing. These branches and a wreath ring which can be bought from a florist wholesaler along with any fancy bows or small plastic bells in fact anything that glitters, you can make lots of wreaths and sell them at a good profit for a small expenditure. You can also sell stands for the trees; the other expenditure will be a netting machine to make the tree more compact for transporting in peoples cars. If you decide to grow Christmas trees this is the way to start, you build up a clientele in the years that the trees are growing then you can start selling your own.

Growing Christmas trees is a long term enterprise and selecting the right type of land is critical, the land needs to be sloping or well drained, if the ground wet or water logged it will slow the growth of the trees. The trees do like rain and cold weather but not wet ground. It is said that Christmas trees will grow in bad ground and it is possibly true but you will get bad trees the better the soil the better the tree. Not all trees will grow in all parts of the country you will have to take advice from the sapling supplier on what will grow where. The other decision to make is what type of trees to grow to suit the pockets of your customers, you will not sell many Noordman firs at £30 in a deprived area or indeed a cheap spruce in a wealthy area, you need to work out the combination of the right tree that will grow in the area and sell in the area.

You will be able to plant over 4000 per acre and they will take between 7 to 10 years to get to a saleable size, regardless of the claims made by the sellers of the saplings that they will be saleable in 5 or 6 years. To grow Christmas trees long term you will need to plant 1 acre a year for 10 years then you would have 4000 trees to sell every year for ever. To buy the saplings you will need to give the supplier a lot of notice normally 6 months at least, the price from each supplier will vary enormously sometimes by up to 50% this is justified by the dear ones saying they have better stock, in my opinion this is rubbish. Beware of any buy back deals that may be offered by the sapling supplier they will offer to buy all of you saleable trees. And give you a price per foot normally 50% of retail omitting to mention this is for grade "A" trees and guess who grades the trees the man who is buying them. Regardless of the quality of the trees only about 20% will be graded "A" the rest will be graded much lower and consequently will command a much lower price, it is always better to avoid wholesaling if at all possible. If you use a pick your own system then you can eliminate waste totally, the client selects the tree and you cut it, if any tree doesn't sell leave it until next year. If you think you will have a surplus then you need to start taking wholesale orders in January for the following December, and the best customers will be independent garages and garden centres. The care and attention that the trees need is minimal, maintaining a good shape for the tree by trimming, and spraying with weed killer as required. I believe that you can get sheep that will graze between the trees without eating them, so as the trees take up more of the land you will not lose all the grazing and you will save on the weed killing spray. If only they had a sheep that could trim the trees as well. I will leave with this thought if you did grow 4000 trees a year and you retailed them at say £25 each that would be £100000 per year.

FARM SHOP.

You are allowed to sell anything at you farm gate / shop that is produced on your land, anything else you will require planning permission for a shop, which is rarely refused if there are no issues with access or parking, the shop will then of course be liable for business rates.

HAY AND STRAW DEALER.

Most farmers now use a round bale system for hay and straw which needs machinery to move them around, this coupled with the rise in the horse population there is a demand for small bales. You buy grass standing and bail it yourself and you can do a deal with local farmers to buy straw from behind the combine and bail it yourself or you can buy it wholesale in the summer from other parts of the country to sell in winter, see if there is a demand in your area. A word of warning when dealing with the horse set always get your money when you give them the hay or straw never give credit or you will never see your money.

TOURING CARAVAN SITE

Certified Locations

Under the "caravan sites and control of development act 1960" certain organisations are exempt from needing planning permission or a site licence and can certify a location to set up a touring caravan site, which means you set up the caravan site and their members use it and pay you. The Caravan Club is one of the exempted organisations and allow certified locations, to be set up for their members use only. The site would be for 5 touring caravans and can be for all of the year. There are no onerous conditions which need to be complied with. However it would be a good idea to put in electrical hook ups to the pitches as it is expected by the users, and a water supply to each pitch would be easy to put in with the electric cable, as in the not to distant future this will also be expected by the users. As was mentioned earlier in the book under access road the gate being moved back 20 metres from the road will improve the safety of the caravans entering and leaving the site.

You will need an ordnance survey map of the area to find out the location of the other touring caravan sites, if there are some locally and it is a high demand area like a tourist destination, or a particular attraction nearby say for bird watching then I am sure the caravan club would agree to your site. If there none in your local area perhaps there is no demand or perhaps no one has wanted to set one up the caravan club with their vast experience can look at a potential site and see if it is viable. Being a caravan club site can be your start into caravans, if you are successful and you are over booked for the season you can apply for planning permission to enlarge the site, and as it is a sustainable tourist related rural business. It is supported by the government particularly with regard to planning permission. When you are a certified location the marketing for your site is done by the Caravan Club, your site would be put in their hand book, in the site directory and on their web site. If you are interested in the idea of a touring caravan site then you need to contact the Caravan Club for an application pack.

SKILLS

You may already posses skills that could be adapted to earn money in an agricultural way or you may like to acquire a skill by way of a course. You could then pass the skill on to other by holding courses, I have listed below a few courses that could be run from the smallholding without planning permission.

Thatching
Dry stone walling
Hedge laying
Hurdle making
Charcoal burning
Bee keeping
Cheese making
Ice cream making
Spinning
Weaving

There are more specialist businesses that can also be undertaken these would require higher levels of skills and knowledge.

Repairing farm machinery
Buying and selling animals
Transporting animals (L G V licence, Operators Licence and planning permission for an operating centre) and of course all the rules and regulation governing animal movements.

LEGISLATION

Dangerous wild animal act 1976 and 1984, this would be applicable in the business listed eg, Wild Boar, Ostriches and some types of camelids which you are unlikely to want to keep, there is an annual licence fee payable to the local authority and also an annual safety and welfare inspection which you must pay for.

If you are retailing then this is a list of the legislation which you operate under it is not exhaustive.

The food safety (general food hygiene) regulations 1995

Food labelling regulations 1996

The weights and measures act 1985

Consumer protection act 1987

Food safety act 1990

CONTENTS

Town and Country Planning (General Permitted Development) order 1995
Part 5 Caravan sites.
Caravan Sites and Control of Development act 1960.(c.62)
These two pieces of legislation should be read together to make any sense.
Town and Country Planning (General Permitted Development) order 1995
Part 6 Agricultural Buildings and Operations.
Polytunnels this gives a brief outline of the court rulings and planning officers' decisions.

The Town and Country Planning (General Permitted Development) Order 1995
CROWN COPYRIGHT

PART 5

CARAVAN SITES

Class A

Permitted development

A. **The use of land, other than a building, as a caravan site in the circumstances referred to in paragraph A.2.**

Condition

A.1 Development is permitted by Class A subject to the condition that the use shall be discontinued when the circumstances specified in paragraph A.2 cease to exist, and all caravans on the site shall be removed as soon as reasonably practicable.

Interpretation of Class A

A.2 The circumstances mentioned in Class A are those specified in paragraphs 2 to 10 of Schedule 1 to the 1960 Act (cases where a caravan site licence is not required), but in relation to those mentioned in paragraph 10 do not include use for winter quarters.

Class B

Permitted development

B. **Development required by the conditions of a site licence for the time being in force under the 1960 Act.**

Caravan Sites and Control of Development Act 1960 Crown Copyright.

FIRST SCHEDULE

CASES WHERE A CARAVAN SITE LICENCE IS NOT REQUIRED

Use within curtilage of a dwellinghouse

1. A site licence shall not be required for the use of land as a caravan site if the use is incidental to the enjoyment as such of a dwellinghouse within the curtilage of which the land is situated.

Use by a person travelling with a caravan for one or two nights

2. Subject to the provisions of paragraph 13 of this Schedule, a site licence shall not be required for the use of land as a caravan site by a person travelling with a caravan who brings the caravan on to the land for a period which includes not more than two nights—

(a) if during that period no other caravan is stationed for the purposes of human habitation on that land or any adjoining land in the same occupation, and

(b) if, in the period of twelve months ending with the day on which the caravan is brought on to the land, the number of days on which a caravan was stationed anywhere on that land or the said adjoining land for the purposes of human habitation did not exceed twenty-eight.

Use of holdings of five acres or more in certain circumstances

3. — (1) Subject to the provisions of paragraph 13 of this Schedule, a site licence shall not be required for the use as a caravan site of land which comprises, together with any adjoining land which is in the same occupation and has not been built on, not less than five acres—

(a) if in the period of twelve months ending with the day on which the land is used as a caravan site the number of days on which a caravan was stationed anywhere on that land or on the said adjoining land for the purposes of human habitation did not exceed twenty-eight, and

(b) if in the said period of twelve months not more than three caravans were so stationed at any one time.

(2) The Minister may by order contained in a statutory instrument provide that in any such area as may be specified in the order this paragraph shall have effect subject to the modification—

(a) that for the reference in the foregoing sub-paragraph to five acres there shall be substituted a reference to such smaller acreage as may be specified in the order, or

(b) that for the condition specified in head (a) of that sub-paragraph there shall be substituted a condition that the use in question falls between such dates in any year as may be specified in the order,

or subject to modification in both such respects.

(3) The Minister may make different orders under this paragraph as respects different areas, and an order under this paragraph may be varied by a subsequent order made thereunder.

(4) An order under this paragraph shall come into force on such date as may be specified in the order, being a date not less than three months after the order is made; and the Minister shall publish notice of the order in a local newspaper circulating in the locality affected by the order and in such other ways as appear to him to be expedient for the purpose of drawing the attention of the public to the order.

Sites occupied and supervised by exempted organisations

4. Subject to the provisions of paragraph 13 of this Schedule, a site licence shall not be required for the use as a caravan site of land which is occupied by an organisation which holds for the time being a certificate of exemption granted under paragraph 12 of this Schedule (hereinafter referred to as an exempted organisation) if the use is for purposes of recreation and is under the supervision of the organisation.

Sites approved by exempted organisations

5. — (1) Subject to the provisions of paragraph 13 of this Schedule, a site licence shall not be required for the use as a caravan site of land as respects which there is in force a certificate issued under this paragraph by an exempted organisation if not more than five caravans are at the time stationed for the purposes of human habitation on the land to which the certificate relates.

(2) For the purposes of this paragraph an exempted organisation may issue as respects any land a certificate stating that the land has been approved by the exempted organisation for use by its members for the purposes of recreation.

(3) The certificate shall be issued to the occupier of the land to which it relates, and the organisation shall send particulars to the Minister of all certificates issued by the organisation under this paragraph.

(4) A certificate issued by an exempted organisation under this paragraph shall specify the date on which it is to come into force and the period for which it is to continue in force, being a period not exceeding one year.

Meetings organised by exempted organisations

6. Subject to the provisions of paragraph 13 of this Schedule, a site licence shall not be required for the use of land as a caravan site if the use is under the supervision of an exempted organisation and is in pursuance of arrangements made by that organisation for a meeting for its members lasting not more than five days.

Agricultural and forestry workers

7. Subject to the provisions of paragraph 13 of this Schedule, a site licence shall not be required for the use as a caravan site of agricultural land for the accommodation during a particular season of a person or persons employed in farming operations on land in the same occupation.

8. Subject to the provisions of paragraph 13 of this Schedule, a site licence shall not be required for the use of land as a caravan site for the accommodation during a particular season of a person or persons employed on land in the same occupation, being land used for the purposes of forestry (including afforestation).

Building and engineering sites

9. Subject to the provisions of paragraph 13 of this Schedule, a site licence shall not be required for the use as a caravan site of land which forms part of, or adjoins, land on which building or engineering operations are being carried out (being operations for the carrying out of which permission under Part III of the Act of 1947 has, if required, been granted) if that use is for the accommodation of a person or persons employed in connection with the said operations.

Travelling showmen

10. — (1) Subject to the provisions of paragraph 13 of this Schedule, a site licence shall not be required for the use of land as a caravan site by a travelling showman who is a member of an organisation of travelling showmen which holds for the time being a certificate granted under this paragraph and who is, at the time, travelling for the purposes of his business or who has taken up winter quarters on the land with his equipment for some period [**F1**falling between the beginning of October in any year and the end of March] [**F1**beginning on or after 20 September in any year and continuing until not later than 16 April] in the following year.

(2) For the purposes of this paragraph the Minister may grant a certificate to any organisation recognised by him as confining its membership to bona fide travelling showmen; and a certificate so granted may be withdrawn by the Minister at any time.

Annotations:

Amendments (Textual)

F1Words "beginning on or after 20 September in any year and continuing until not later than 16 April" substituted (S.) for words "falling between the beginning of October in any year and the end of March" by Local Government and Planning (Scotland) Act 1982 (c. 43, SIF 81:2), s. 66(1), Sch. 3 para. 3(*a*)

Sites occupied by licensing authority

11. A site licence shall not be required for the use as a caravan site of land occupied by the local authority in whose area the land is situated.

Annotations:

Modifications etc. (not altering text)

C1Sch. 1 para. 11 extended (E.W.) (19.9.1995) by 1995 c. 25, ss. 70, 125(2), Sch. 9 para. 4(b) (with ss. 7(6), 115, 117, Sch. 8 para. 7)

[**F2***Gipsy sites occupied by county councils or regional councils*

Annotations:

Amendments (Textual)

F2Para. 11A inserted by Local Government, Planning and Land Act 1980 (c. 65, SIF 81:1, 2), s. 176

[**F3**11A. A site licence shall not be required for the use of land occupied by a county council, or in Scotland by a regional council, as a caravan site providing accommodation for [**F4**persons to whom section 24(8A) of this Act applies].]

Annotations:

Amendments (Textual)

F3Para. 11A inserted by Local Government, Planning and Land Act 1980 (c. 65, SIF 81:1, 2), s. 176

F4Words substituted (S.) by virtue of Local Government and Planning (Scotland) Act 1982 (c.43, SIF 81:2), s.66(1), Sch. 3 para. 3(*b*)

Certification of exempted organisations

12. — (1) For the purposes of paragraphs 4, 5 and 6 of this Schedule the Minister may grant a certificate of exemption to any organisation as to which he is satisfied that it objects include the encouragement or promotion of recreational activities.

(2) A certificate granted under this paragraph may be withdrawn by the Minister at any time.

The Town and Country Planning (General Permitted Development) order1995 Crown Copyright.

PART 6

AGRICULTURAL BUILDINGS AND OPERATIONS

Class A Development on units of 5 hectares or more
Permitted development
A. The carrying out on agricultural land comprised in an agricultural unit of 5 hectares or more in area of—

(a) works for the erection, extension or alteration of a building; or

(b) any excavation or engineering operations,

which are reasonably necessary for the purposes of agriculture within that unit.
Development not permitted
A.1 Development is not permitted by Class A if—

(a) the development would be carried out on a separate parcel of land forming part of the unit which is less than 1 hectare in area;

(b) it would consist of, or include, the erection, extension or alteration of a dwelling;

(c) it would involve the provision of a building, structure or works not designed for agricultural purposes;

(d) the ground area which would be covered by—

(i) any works or structure (other than a fence) for accommodating livestock or any plant or machinery arising from engineering operations; or

(ii) any building erected or extended or altered by virtue of Class A, would exceed 465 square metres, calculated as described in paragraph D.2 below;

(e) the height of any part of any building, structure or works within 3 kilometres of the perimeter of an aerodrome would exceed 3 metres;

(f) the height of any part of any building, structure or works not within 3 kilometres of the perimeter of an aerodrome would exceed 12 metres;

(g) any part of the development would be within 25 metres of a metalled part of a trunk road or classified road;

(h) it would consist of, or include, the erection or construction of, or the carrying out of any works to, a building, structure or an excavation used or to be used for the accommodation of livestock or for the storage of slurry or sewage sludge where the building, structure or excavation is, or would be, within 400 metres of the curtilage of a protected building; or

(i) it would involve excavations or engineering operations on or over article 1(6) land which are connected with fish farming.

Conditions

A.2(1) Development is permitted by Class A subject to the following conditions—

(a) where development is carried out within 400 metres of the curtilage of a protected building, any building, structure, excavation or works resulting from the development shall not be used for the accommodation of livestock except in the circumstances described in paragraph D.3 below or for the storage of slurry or sewage sludge;

(b) where the development involves—

(i) the extraction of any mineral from the land (including removal from any disused railway embankment); or

(ii) the removal of any mineral from a mineral-working deposit, the mineral shall not be moved off the unit;

(c) waste materials shall not be brought on to the land from elsewhere for deposit except for use in works described in Class A(a) or in the provision of a hard surface and any materials so brought shall be incorporated forthwith into the building or works in question.

(2) Subject to paragraph (3), development consisting of—

(a) the erection, extension or alteration of a building;

(b) the formation or alteration of a private way;

(c) the carrying out of excavations or the deposit of waste material (where the relevant area, as defined in paragraph D.4 below, exceeds 0.5 hectare); or

(d) the placing or assembly of a tank in any waters,

is permitted by Class A subject to the following conditions—

(i) the developer shall, before beginning the development, apply to the local planning authority for a determination as to whether the prior approval of the authority will be required to the siting, design and external appearance of the building, the siting and means of construction of the private way, the siting of the excavation or deposit or the siting and appearance of the tank, as the case may be;

(ii) the application shall be accompanied by a written description of the proposed development and of the materials to be used and a plan indicating the site together with any fee required to be paid;

(iii) the development shall not be begun before the occurrence of one of the following—

    (aa) the receipt by the applicant from the local planning authority of a written notice of their determination that such prior approval is not required;

    (bb) where the local planning authority give the applicant notice within 28 days following the date of receiving his application of their determination that such prior approval is required, the giving of such approval; or

    (cc) the expiry of 28 days following the date on which the application was received by the local planning authority without the local planning authority making any determination as to whether such approval is required or notifying the applicant of their determination;

(iv)

    (aa) where the local planning authority give the applicant notice that such prior approval is required the applicant shall display a site notice by site display on or near the land on which the proposed development is to be carried out, leaving the notice in position for not less than 21 days in the period of 28 days from the date on which the local planning authority gave the notice to the applicant;

    (bb) where the site notice is, without any fault or intention of the applicant, removed, obscured or defaced before the period of 21 days referred to in sub-paragraph (aa) has elapsed, he shall be treated as having complied with the requirements of that sub-paragraph if he has taken reasonable steps for protection of the notice and, if need be, its replacement;

(v) the development shall, except to the extent that the local planning authority otherwise agree in writing, be carried out—

(aa) where prior approval is required, in accordance with the details approved;

(bb) where prior approval is not required, in accordance with the details submitted with the application; and

(vi) the development shall be carried out—

(aa) where approval has been given by the local planning authority, within a period of five years from the date on which approval was given;

(bb) in any other case, within a period of five years from the date on which the local planning authority were given the information referred to in sub-paragraph (d)(ii).

(3) The conditions in paragraph (2) do not apply to the extension or alteration of a building if the building is not on article 1(6) land except in the case of a significant extension or a significant alteration.

(4) Development consisting of the significant extension or the significant alteration of a building may only be carried out once by virtue of Class A(a).

Class B Development on units of less than 5 hectares

Permitted development

B. The carrying out on agricultural land comprised in an agricultural unit of not less than 0.4 but less than 5 hectares in area of development consisting of—

(a) the extension or alteration of an agricultural building;

(b) the installation of additional or replacement plant or machinery;

(c) the provision, rearrangement or replacement of a sewer, main, pipe, cable or other apparatus;

(d) the provision, rearrangement or replacement of a private way;

(e) the provision of a hard surface;

(f) the deposit of waste; or

(g) the carrying out of any of the following operations in connection with fish farming, namely, repairing ponds and raceways; the installation of grading machinery, aeration equipment or flow meters and any associated channel; the dredging of ponds; and the replacement of tanks and nets,

where the development is reasonably necessary for the purposes of agriculture within the unit.

Development not permitted

B.1 Development is not permitted by Class B if—

(a) the development would be carried out on a separate parcel of land forming part of the unit which is less than 0.4 hectare in area;

(b) the external appearance of the premises would be materially affected;

(c) any part of the development would be within 25 metres of a metalled part of a trunk road or classified road;

(d) it would consist of, or involve, the carrying out of any works to a building or structure used or to be used for the accommodation of livestock or the storage of slurry or sewage sludge where the building or structure is within 400 metres of the curtilage of a protected building; or

(e) it would relate to fish farming and would involve the placing or assembly of a tank on land or in any waters or the construction of a pond in which fish may be kept or an increase (otherwise than by the removal of silt) in the size of any tank or pond in which fish may be kept.

B.2 Development is not permitted by Class B(a) if—

(a) the height of any building would be increased;

(b) the cubic content of the original building would be increased by more than 10%;

(c) any part of any new building would be more than 30 metres from the original building;

(d) the development would involve the extension, alteration or provision of a dwelling;

(e) any part of the development would be carried out within 5 metres of any boundary of the unit; or

(f) the ground area of any building extended by virtue of Class B(a) would exceed 465 square metres.

B.3 Development is not permitted by Class B(b) if—

(a) the height of any additional plant or machinery within 3 kilometres of the perimeter of an aerodrome would exceed 3 metres;

(b) the height of any additional plant or machinery not within 3 kilometres of the perimeter of an aerodrome would exceed 12 metres;

(c) the height of any replacement plant or machinery would exceed that of the plant or machinery being replaced; or

(d) the area to be covered by the development would exceed 465 square metres calculated as described in paragraph D.2 below.

B.4 Development is not permitted by Class B(e) if the area to be covered by the development would exceed 465 square metres calculated as described in paragraph D.2 below.

Conditions

B.5 Development permitted by Class B and carried out within 400 metres of the curtilage of a protected building is subject to the condition that any building which is extended or altered, or any works resulting from the development, shall not be used for the accommodation of livestock except in the circumstances described in paragraph D.3 below or for the storage of slurry or sewage sludge.

B.6 Development consisting of the extension or alteration of a building situated on article 1(6) land or the provision, rearrangement or replacement of a private way on such land is permitted subject to—

(a) the condition that the developer shall, before beginning the development, apply to the local planning authority for a determination as to whether the prior approval of the authority will be required to the siting, design and external appearance of the building as extended or altered or the siting and means of construction of the private way; and

(b) the conditions set out in paragraphs A.2(2)(ii) to (vi) above.

B.7 Development is permitted by Class B(f) subject to the following conditions—

(a) that waste materials are not brought on to the land from elsewhere for deposit unless they are for use in works described in Class B(a), (d) or (e) and are incorporated forthwith into the building or works in question; and

(b) that the height of the surface of the land will not be materially increased by the deposit.

Class C Mineral working for agricultural purposes
Permitted development

C. The winning and working on land held or occupied with land used for the purposes of agriculture of any minerals reasonably necessary for agricultural purposes within the agricultural unit of which it forms part.

Development not permitted

C.1 Development is not permitted by Class C if any excavation would be made within 25 metres of a metalled part of a trunk road or classified road.

Condition

C.2 Development is permitted by Class C subject to the condition that no mineral extracted during the course of the operation shall be moved to any place outside the land from which it was extracted, except to land which is held or occupied with that land and is used for the purposes of agriculture.

Interpretation of Part 6

D.1 For the purposes of Part 6—

"agricultural land" means land which, before development permitted by this Part is carried out, is land in use for agriculture and which is so used for the purposes of a trade or business, and excludes any dwellinghouse or garden;

"agricultural unit" means agricultural land which is occupied as a unit for the purposes of agriculture, including—

(a) any dwelling or other building on that land occupied for the purpose of farming the land by the person who occupies the unit, or

(b) any dwelling on that land occupied by a farmworker;

"building" does not include anything resulting from engineering operations;

"fish farming" means the breeding, rearing or keeping of fish or shellfish (which includes any kind of crustacean and mollusc);

"livestock" includes fish or shellfish which are farmed;

"protected building" means any permanent building which is normally occupied by people or would be so occupied, if it were in use for purposes for which it is apt; but does not include—

(i) a building within the agricultural unit; or

(ii) a dwelling or other building on another agricultural unit which is used for or in connection with agriculture;

"significant extension" and "significant alteration" mean any extension or alteration of the building where the cubic content of the original building would be exceeded by more than 10% or the height of the building as extended or altered would exceed the height of the original building;

"slurry" means animal faeces and urine (whether or not water has been added for handling); and

"tank" includes any cage and any other structure for use in fish farming.

D.2 For the purposes of Part 6—

(a) an area calculated as described in this paragraph comprises the ground area which would be covered by the proposed development, together with the ground area of any building (other than a dwelling), or any structure, works, plant, machinery, ponds or tanks within the same unit which are being provided or have been provided within the preceding two years and any part of which would be within 90 metres of the proposed development;

(b) 400 metres is to be measured along the ground.

D.3 The circumstances referred to in paragraphs A.2(1)(a) and B.5 are—

(a) that no other suitable building or structure, 400 metres or more from the curtilage of a protected building, is available to accommodate the livestock; and

(b)

    (i) that the need to accommodate the livestock arises from—

        (aa) quarantine requirements; or

        (bb) an emergency due to another building or structure in which the livestock could otherwise be accommodated being unavailable because it has been damaged or destroyed by fire, flood or storm; or

    (ii) in the case of animals normally kept out of doors, they require temporary accommodation in a building or other structure—

        (aa) because they are sick or giving birth or newly born; or

        (bb) to provide shelter against extreme weather conditions.

D.4 For the purposes of paragraph A.2(2)(c), the relevant area is the area of the proposed excavation or the area on which it is proposed to deposit waste together with the aggregate of the areas of all other excavations within the unit which have not been filled and of all other parts of the unit on or under which waste has been deposited and has not been removed.

D.5 In paragraph A.2(2)(iv), "site notice" means a notice containing—

(a) the name of the applicant,

(b) the address or location of the proposed development,

(c) a description of the proposed development and of the materials to be used,

(d) a statement that the prior approval of the authority will be required to the siting, design and external appearance of the building, the siting and means of construction of the private way, the siting of the excavation or deposit or the siting and appearance of the tank, as the case may be,

(e) the name and address of the local planning authority,

and which is signed and dated by or on behalf of the applicant.

D.6 For the purposes of Class B—

(a) the erection of any additional building within the curtilage of another building is to be treated as the extension of that building and the additional building is not to be treated as an original building;

(b) where two or more original buildings are within the same curtilage and are used for the same undertaking they are to be treated as a single original building in making any measurement in connection with the extension or alteration of either of them.

D.7 In Class C, "the purposes of agriculture" includes fertilising land used for the purposes of agriculture and the maintenance, improvement or alteration of any buildings, structures or works occupied or used for such purposes on land so used.

Power to withdraw certain exemptions

13. — (1) The Minister may on the application of a local authority by order provide that, in relation to such land situated in their area as may be specified in the order, this Schedule shall have effect as if paragraphs 2 to 10, or such one or more of those paragraphs as may be so specified, were omitted from this Schedule.

(2) An order under this paragraph—

(a)

shall come into force on such date as may be specified therein, and

(b)

may, on the application of the local authority on whose application it was made, be varied or revoked by a subsequent order made thereunder,

and, except in the case of an order the sole effect of which is to revoke in whole or part a previous order, the local authority shall, not less than three months before the order comes into force, cause a notice setting out the effect of the order and the date on which it comes into force to be published in the London Gazette or, if the land is in Scotland, in the Edinburgh Gazette and in a local newspaper circulating in the locality in which the land to which the order relates is situated.